IDEAS OF SCIENCE

IDEAS OF SCIENCE

Bernard Dixon
Geoffrey Holister

Basil Blackwell

Published by Basil Blackwell Limited
108 Cowley Road
Oxford OX4 1JF
England

Typeset by Katerprint Co Ltd, Oxford

Printed by Dotesios (Printers) Ltd. Bradford-on-Avon, Wiltshire

British Library Cataloguing in Publication Data

Holister, Geoffrey
  Ideas of science.
  1. Science – Social aspects
  I. Title   II. Dixon, Bernard
  306'.45   Q175.5

ISBN 0 631 13967 2

Contents

Acknowledgements		*vi*
1	Science: for whom and for what?	1
2	Cause and effect	5
3	Modelling	8
4	Reproducibility	11
5	Controls	14
6	Science and certainty	17
7	Statistics	20
8	Science and reality	23
9	Theory and experiment in science	27
10	Thought experiments	32
11	Science and imagination	35
12	Science and pseudo-science	39
13	Science and human values	47
14	Science and social responsibility	51
	Further reading	54

Acknowledgements

Our thanks are due to Paul Holister for Chapter 10 'Thought Experiments'.

The authors and publishers wish to acknowledge the following for permission to reproduce illustrations:

Abercrombie's model of a depression (from Sutton, O.G., *Understanding Weather*, Pelican Books, third edition 1969, p. 82), reproduced by permission of Penguin Books Ltd   9 (Figure 5)

David Austin   1, 23, 53

British Broadcasting Corporation   9 (Figure 3)

Cambridge Instruments Ltd   cover, 30

L. D. Harmon for the illustration of pattern recognition from Harmon, L.D., and Julesz, B., 'Masking in Visual Recognition Effects of Two-Dimensional Filtered Noise', reproduced in *Science*, volume 180, 15 June 1983, cover, copyright 1983 by the AAAS   36 (Figure 2)

Merrily Harpur/*The Guardian*   19 (Figure 1)

London Transport (Underground map, registered user number 84/237)   8 (Figure 1)

Mansell Collection   32 (Figure 1)

Hilary Roberts, from 'How to teach scientists to start worrying and hate the bomb', *The Guardian*, 1 February 1984   52

Illustrations by Tim Smith   4, 5, 8 (Figure 2), 13, 20, 22, 29 (Figure 2), 33 (Figure 2), 35 (Figure 1), 36 (Figure 3), 42 (Figure 2)

The Hidden Player at Life's Game

Suppose it were perfectly certain that the life and fortune of every one of us would, one day or the other, depend upon his winning or losing a game of chess. Don't you think that we should all consider it to be a primary duty to learn at least the names and the moves of the pieces; to have a notion of a gambit, and a keen eye for all the means of giving and getting out of check? Do you not think that we should look with a disapprobation amounting to scorn, upon the father who allowed his son, or the state which allowed its members, to grow up without knowing a pawn from a knight?

Yet it is a very plain and elementary truth, that the life, the fortune, and the happiness of every one of us, and, more or less, of those who are connected with us, do depend upon our knowing something of the rules of a game infinitely more difficult and complicated than chess. It is a game which has been played for untold ages, every man and woman of us being one of the two players in a game of his or her own. The chessboard is the world, the pieces are the phenomena of the universe, the rules of the game are what we call the laws of Nature. The player on the other side is hidden from us. We know that his play is always fair, just and patient. But also we know, to our cost, that he never overlooks a mistake, or makes the smallest allowance for ignorance. To the man who plays well, the highest stakes are paid, with that sort of overflowing generosity with which the strong shows delight in strength. And one who plays ill is checkmated – without haste, but without remorse.

T. H. Huxley, 'The Hidden Player at Life's Game', from 'A liberal education; and where to find it', *Macmillan's Magazine*, March 1868, **17**: 367.

1 Science: for whom and for what?

'Art is "I", science is "we"'   anon.

Two caricatures of the scientist are irritatingly persistent in the media; both are male and white-coated – one tidy-minded and boring, the other maniacal and eccentric. Anyone studying science or contemplating a career in science should forget both images – and not just because women's achievements make a masculine archetype absurd. The other reason for rejecting such pictures is because, however much they contrast in other ways, they perpetuate another myth: that science is a uniform, predictable and monotonous activity. Both the sober boffin and the mad scientist are stereotypes which obscure the great variety of scientific life.

The mixture of speakers assembled for the 1982 meeting of the British Association for the Advancement of Science illustrated the point beautifully. There was James Lovelock who, while holidaying in Eire, first wondered whether the fluorocarbons used as aerosol propellants were accumulating in the upper atmosphere. He worked out that they probably were, and went on to alert us to the possibility that they might be depleting the earth's ozone layer. Lovelock is very much an individualist, he carries out all sorts of research projects in a laboratory at home and travels around the world to discuss his findings with other scientists. At the other extreme, the BA meeting heard from Chris Damarell, a high-energy physicist who works with particle accelerators such as those at CERN in Geneva. He works as part of a team which is so large that one member may not even meet all the others. Managerial skills are vitally important for someone leading a team like this – the ability to bring together the talents of many different scientists and indeed types of scientist, perhaps based around the world.

Just as revealing were the contributors to the conference who talked about the way their research had *really* been done – something that is often poorly conveyed in the learned journals where scientists report their findings. On the one hand there were happy accidents, illustrated by

the laboratory assistant's mistake in formulating a culture medium, which led to the first isolation of the viruses responsible for the common cold. On the other hand, a more Sherlock Holmesean brand of research was represented by two tales of detection, one from the physical sciences and the other from biology. First Jocelyn Bell, who discovered the peculiar, pulsating radio stars now known as pulsars, described the excitement of the moment and the mountains of scrupulous toil need to clarify the significance of the bizarre radio signals when they were first received in Cambridge. Then E. S. Anderson, who masterminded investigations into the typhoid fever outbreak in Aberdeen in 1964, and many similar epidemics, talked about a type of scientific sleuthing that is much nearer to the work of a fictional detective than that of a fictional boffin.

Another contrast was provided by Martin Rees, a theoretical physicist who rarely or never goes near a laboratory bench. He is interested in the energy transformations in objects such as pulsars, and although he uses measurements and signals reported by radio astronomers and other types of experimenter, he does not conduct experiments himself. Theoretical physics, by definition, advances by computation and by mathematical exploration of hypotheses rather than by practical dexterity with laboratory equipment.

Biological, physical and social sciences; pure research and applied research; dogged computation and inspired guesswork; teamwork and individual talent; spectacular breakthroughs and 'normal science' that proceeds methodically between periods of revolution – these are just some of the different faces of science. Think of everyday life and others immediately come to mind. Tracking down a food poisoning epidemic is an exciting challenge, but there is considerable scientific skill and dedication in the public health work which, day by day, ensures that Britain's cities are free of food poisoning, typhoid fever and dysentery. The investigation of air accidents requires collaboration between many different sorts of scientist and engineer. A similar input of skill is needed, 24 hours in every day, to maintain the excellent safety record of the world's airlines.

In his fine book *Advice to a Young Scientist*, Sir Peter Medawar says: 'Among scientists are collectors, classifiers, and compulsive tidiers-up; many are detectives by temperament and many are explorers; some are artists and others artisans. There are poet-scientists and philosopher-scientists, and even a few mystics'. This list is probably incomplete but it does illustrate the remarkable spectrum of activities encompassed by that one term 'scientist'. Science is not a monolithic enterprise but a galaxy of different activities, providing job satisfaction and excitement to people of contrasting abilities and personalities.

Although this book is called *Ideas of Science*, it is also aimed at students of engineering. After all, the practice of engineering and the ideas of science are different sides of the same coin, and it is a surprising fact that the study of engineering and the applied sciences does not have the prestige attached to it in this country that it does in other developed countries.

The concept of the engineer as a sort of high class tinker who picks over the discoveries of science in order to devise better machines is historically incorrect. The telescope was invented before the laws of refraction were properly understood: it was the need to develop better telescopes that led to the study of optics by Newton and others. Steam engines were developed before the laws of thermodynamics were understood, indeed the second law of thermodynamics arose directly from Watt's efforts to make his steam engines more efficient. The first aeroplane flew before the laws of aerodynamics were developed; and so on. Scientific discovery and engineering invention are inextricably linked, sometimes one leading, sometimes the other; the nuclear power station is an example of science leading technology.

In recent years many concepts, developed as a result of new technologies, have been responsible for exciting new insights in science. An example of this is the discovery, through efforts to increase the message-carrying capacity of telephone cables and radio channels, of *information* as a scientifically useful concept, and in particular the basic quantum of information, the BIT (short for 'binary digit', or Yes/No statement). Other fascinating scientific insights, particularly in biology, may result from engineering studies of pattern recognition and other aspects associated with 'intelligent' computers.

Engineering is an integral part of science. Of course there is a difference in what is expected of engineers and scientists. If a scientist is working on the structure of DNA and he accidentally discovers a cure for cancer he will probably get a Nobel Prize. On the other hand, if an engineer is designing a gas turbine and accidentally develops a new type of vacuum cleaner he will probably get fired.

However variegated the scientific community, there *is* something that binds all scientists together – a certain attitude of mind, an acute regard for intellectual rigour, a critical stance towards evidence, and an insistence on due caution when conclusions are drawn. In this book, we shall be exploring key aspects of this approach, sometimes called the 'scientific method', which underlies the rich variety of occupations and preoccupations we have discussed so far. We shall see that scientific method is not something to be reserved for use inside laboratory blocks and research institutes; as scientific issues impinge ever more closely on society at large, scientific method is important to every one of us as a private citizen.

"Up to this point, his logic is perfect."

2 Cause and effect

Imagine an extra-terrestrial visitor, arriving on Earth to compare our laws and our regard for them with what is found on other planets. He, she or it spends six months in London and notices sporadic criminal acts – burglaries, occasional murders, some violence in the streets. Our alien also sees the police going about their work – sometimes singly, sometimes in twos or threes. After monitoring the scene for six months or so the visitor comes to a conclusion – the police are the cause of crime. The evidence is simple and obvious: the police are invariably found at the scene of criminal acts. The relationship is too close to be a result of chance. While the majority of British citizens behave quite lawfully, police officers are disproportionately associated with crimes.

With the information *we* have as earthlings, we know at once that this deduction is absurd. But do not be deceived into thinking that our alien's conclusion was obviously, self-evidently crazy. All of us – and that includes politicians, journalists, doctors, and even scientists – are in continual danger of making the same type of mistake. The beguilingly-simple nature of the extra-terrestrial's fallacy, once pointed out, does not prevent other people from falling into the same trap, without a glimmer of suspicion.

Consider two examples of newspaper stories – both fictional and both typical of the type of fallacious reasoning we are exposed to daily. First: drinking lots of tea during pregnancy makes your skin very dry. Second: a vigorously growing potato crop encourages *Patulina* fungus to colonise the soil. In each case, there is impressive evidence to support the claim. Most pregnant women who consume lots of tea develop a dry, scaly skin, while those who drink little or no tea suffer no such symptoms. Gardeners have noticed that thriving potato plants attract *Patulina* to the garden but mediocre or poor plants do not.

The evidence is impressive but by no means conclusive. In the first case, heavy tea drinking *might* lead to dryness of the skin but it is equally possible that there is simply a close overlap between women who crave tea when pregnant and those who develop a dry skin when pregnant. In other words, both outward signs are results of pregnancy. There is no direct, causal link between them.

Likewise with the potato crops. Lustily-growing plants *might* generate nutrients which attract the fungus and prompt it to thrive. However precisely the reverse explanation could be true. Perhaps this particular fungus exudes a nutrient which makes potatoes flourish.

In neither case can we draw a firm conclusion from the available evidence. Further observations or experiments are necessary. For example, a group of pregnant women who would otherwise avoid tea entirely might volunteer to drink lots of it. If the original deduction is correct, then their skin will become dry. If the alternative deduction is correct, there will be no such change. Similar tests with potatoes and *Patulina* fungus could establish the truth in that case.

The basic fallacy is this. When you observe a close, perhaps invariable, relationship between A and B, it may seem inescapably compelling for you to reach the commonsense verdict that A causes B. What you *ought* to do, is to consider two rival explanations – that B causes A, or that both A and B are caused by a third (perhaps unknown) factor, C.

The relationship between lung cancer and smoking is a classic example. In the 1950s it became clear that there was a strikingly-close relationship between this disease and regular smoking. People who smoked cigarettes heavily appeared much more likely to develop lung cancer than those who smoked very little or not at all. The commonsense conclusion was clear. But it could have been utterly wrong. Perhaps the smoking habit and a propensity to develop lung cancer resulted from a third, inbred factor. If that were true, anti-tobacco propaganda would have been both ill-founded and ineffective.

It took many years of research – including studies on different populations, on the same populations when smoking habits were changing, and on the altered prevalence of lung cancer among groups of smokers who gave up compared with those who did not – before it became certain that the 'obvious' conclusion in this case was correct. Smoking does indeed encourage the development of lung cancer.

However commonsense cannot always be trusted. Alongside this success story of medical understanding, there have been abject failures. Microbiologists, for example, have found particular microbes in an impressively high proportion of patients suffering from conditions such as cancer, rheumatoid arthritis or multiple sclerosis, and have deduced accordingly that the organisms must cause the disease. These claims have not stood the test of time. In some cases, the reverse explanation has proved to be true: organisms have infected individuals *because* they were debilitated already. Causality is one of the fundamental concepts of science and needs to be learned and re-learned.

Monday evening I drank whisky and water . . .

Wednesday, I tried brandy and water . . .

. . . and got drunk.

. . . with the same result.

So, on Tuesday I switched to gin and water . . .

The conclusion was obvious: the common factor was water . . .

. . . and got drunk.

. . . which I have therefore decided to give up.

3 Modelling

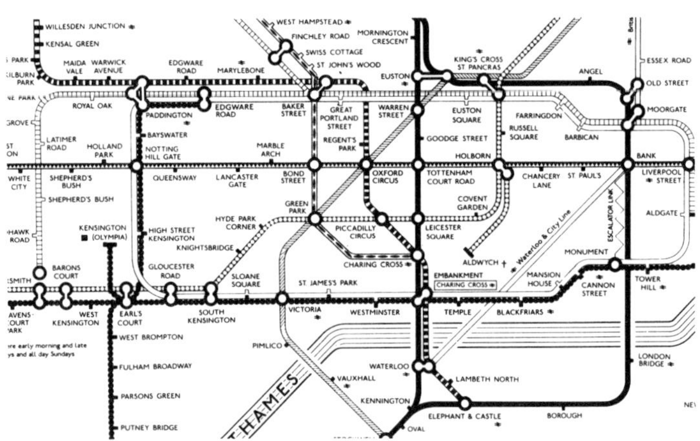

Figure 1

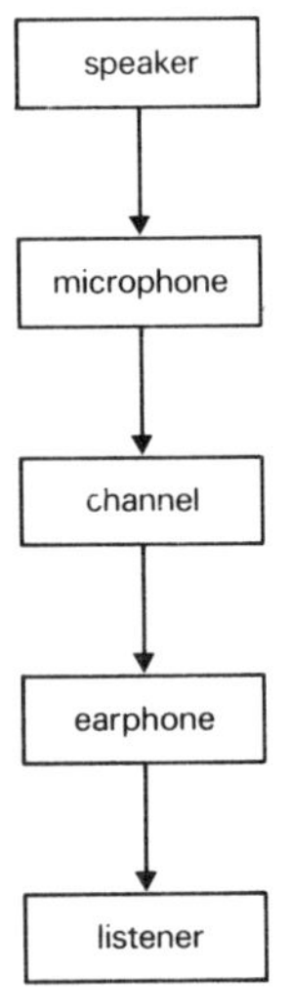

Figure 2

The word modelling, to most people, means children's toys and model aeroplanes, or, at the other extreme, slender girls demonstrating exotic clothes. To scientists and engineers, however, the word modelling describes a way of thinking about the world which is at the heart of the scientific method. In science and technology we make models of things in order to understand them better. Such models, however, do not consist only of physical models made of wood, metal and plastics: scientific models are a simplified representation of reality, created for a specific purpose. Models in this sense can be symbolic, diagrammatic or mathematical. For example, the map of the London Underground (Fig. 1) is a model which simplifies and distorts the true geographical layout in order to represent clearly the routes to be taken between any two points. The block diagram of a telephone circuit (Fig. 2) is a systems model which omits any description of the technical components in order to make clear the fundamental purposes of the various parts of the system as a whole. Similarly, the model audience used for testing the acoustic properties of a scaled-down model of an auditorium (Fig. 3) is constructed to represent the purely acoustical effect of human bodies on sound waves – they do not need to look like real people in order to serve their purpose. The scale model of an aeroplane (Fig. 4), on the other hand, does look like the original because its purpose is to determine the aerodynamic characteristics of the prototype.

These last two examples are typical engineering models and their primary purpose is to predict particular characteristics before time and money are spent building full-scale prototypes. Such models are needed in order to reduce a complex problem to a series of simpler, more manageable problems (Fig. 5). In science, as distinct from technology, modelling is used primarily as a means of explaining some aspect of the physical world. A scientific theory is also a model but because such theories always simplify reality, it follows that the real world seldom corresponds exactly with the predictions of even a very good model.

Often a model is very good under some conditions and very bad under others. For example, the model of the relationship between the pressure, volume and temperature of a mass of gas is very good if the temperature is high enough. At temperatures approaching those at which the gas could condense as a liquid, the model is less accurate, its accuracy deteriorating with decreasing temperature until, rather than calling the model inaccurate, we would call it invalid and look for a different one.

Scientists build up their picture of the cosmos through such models, which does not mean to say that they 'believe' in their models in the sense that a religious person uses the word belief. Indeed, it is the essentially provisional nature of the scientific view that is so difficult to transmit to the non-scientist; particularly one holding a view of the universe as 'an article of faith'. It is a prevalent misunderstanding which is at the heart of much unnecessary conflict between science and religion.

However it is not only scientists and engineers who use models as a means of clarifying their thinking – everybody does it. The words we use are, in a very real sense, models of reality; and as is true of all models, words simplify to some degree the reality they represent. The usefulness of words lies in this simplification: human beings can only function effectively in a complex world by mentally imposing some form of simplifying order on it. This was recognised by the philosopher Wittgenstein, who wrote in his *Tractatus Logico – Philosophicus*:

Figure 3

Figure 4   A scale model of Concorde

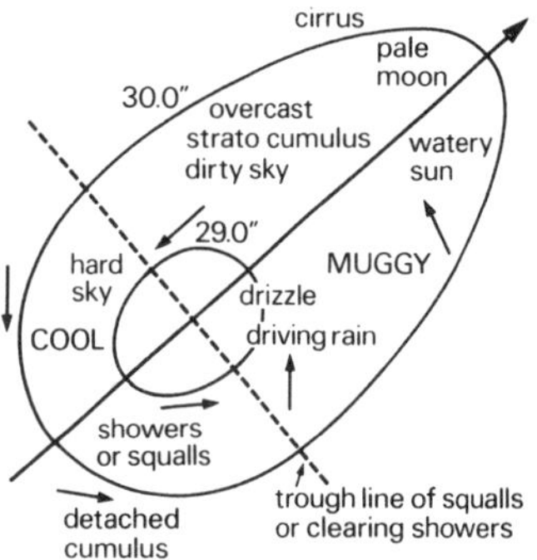

Figure 5   A model of a depression

* The structure of the fact consists of the atomic facts.
* We make to ourselves pictures of facts.
* The picture is a model of reality.

So the words we use and even the thoughts we think are, in this sense, models.

The past few years have seen one of the most ambitious programmes to model our world ever conceived, initiated by Forrester and Meadows (Professors of Engineering at the Massachusetts Institute of Technology). Their purpose was to build a computer model of the world's environmental and economic condition. In order to construct such a model they fed into a computer vast quantities of data concerning such matters as mineral resource depletion, industrial investment, pollution, population growth and so on. Their results were described in the now famous report on the first study of the Club of Rome and its sequel (*The Limits to Growth* and *Mankind at the Turning Point*, see p. 54, *Further Reading List*). These publications caused a furore in scientific and economic circles, and there was considerable criticism of the authors for even attempting to model anything as complex as the world's economic and environmental condition. Forrester responded to this criticism with the following argument:

> A lot of people believe this [building of models] is impossible, because they feel that one cannot make a model of anything as complicated as a social system, and of course they are partially correct – a model is a great simplification. But we have no choice about the use of models. The mental image that we use for passing laws, for running cities, for operating a government – these mental images are models – because one does not have a city, or a country, or a world in one's head, one only has certain images – which are models – and so the question now is: Is the model adequate? Is it the *best* model which we can make? – and the answer is no! it is *not* the best model which we can make!
>
> Professor Forrester, interview for Open University programme 'Limits to Growth', 1981

It is difficult to refute this argument, and as computers become more and more powerful and flexible there is little doubt that computer modelling of the complex social, economic, industrial and military problems of the future will become more commonplace.

4 Reproducibility

'. . . play it again, Sam'   *anon.*

One of the most irritating habits of scientists, when they are interviewed on television or in the press, is their refusal to give straight answers to apparently straight questions. They tend to talk about 'preliminary findings', and to emphasise that a particular discovery 'has not yet been confirmed'. All of this is very frustrating for viewers and readers who simply want to hear the good news about a new antibiotic, cancer treatment, or quasar, and who do not wish to be confused by speakers who seem to be far, far too cautious and guarded. After all, these same individuals would be perfectly happy to say honestly what they think about a film, or to voice dogmatically strong opinions about politics. Why is science so different? Why, even when a researcher has completed an original and successful series of experiments, is he or she likely to be less than 100 per cent confident about the precise meaning of the results?

One crucial answer lies in the fact that science, if it is to advance, is essentially a social activity. The image of solitary thinkers, withdrawn from the world, is a popular one. We think of Isaac Newton, alone in a country orchard, contemplating gravity. We recall Albert Einstein, standing on a bridge over a river and seeing in the water the reflection of curved space, while crowds pass by unnoticed. Science does attract solitary minds, and creativity is often a lonely business but the image is still a misleading, incomplete one. The truth is that the ideas of a Newton or an Einstein, the discoveries and inventions of isolated investigators, have to be tried and tested by the scientific community if they are to be approved. The lifeblood of real science is open, rigorous criticism. By the same token monastic science, conducted by individuals cut off from their peers, represents a dead-end. Work carried out in such a situation – however brilliant in other ways – must atrophy and die.

The crucial point is that an experiment is a *reproducible* observation. Its essence is that the experience must be recorded under specified conditions, which are noted in detail, so that it can be repeated both by the original investigator and by others. There are two essential reasons for this. First, because scientists are human, it would be surprising if cheating did not happen from time to time. Just as there is dishonesty in business, so examples come to light occasionally of research workers who have reported fraudulent data. Exposure usually occurs when other scientists consistently fail to reproduce the same findings.

Secondly, even a person of unquestioned integrity may be mistaken and may, for example, misinterpret meter readings or reach wrong conclusions. Consider a biochemist who is studying the way in which potato tubers produce vitamin C. He may believe, after conducting an extensive and scrupulous series of tests, that he has shown the pH (acidity) of the soil to be the major factor in regulating the manufacture of vitamin C. So he publishes his results, together with details of the experiments, and records his conclusions.

His observations interest two other research teams who decide to study the same phenomenon and try to find out *how* soil pH determines the synthesis of vitamin C. However from the outset they fail to confirm the earlier results. They, in turn, publish their negative findings. The three groups then get together, compare notes, and eventually realise that the real factor responsible for the first set of observations was a mineral micronutrient, whose concentration paralleled the acidity of that soil but that mineral was absent from the soil used by the other two groups. The original biochemist had simply, but quite honestly, made the wrong deduction. The important point is that his error may never have come to light had others not set out to reproduce his work.

There are many real examples of this sort from the world of science. One concerns gravity waves, whose existence is postulated by Einstein's general theory of relativity. The question is: can we actually detect them? A gravity wave passing through space is supposed to generate a transient curvature, within which it should be detectable by its effects on the motion of mechanical systems. The difficulty lies in devising equipment sufficiently sensitive to register such short-lived, minute effects. One person, and only one person, has claimed success. During the late 1960s, Professor Joseph Weber, at the University of Maryland, USA, reported that he had developed apparatus capable of recording gravity waves from the centre of the galaxy. However everyone else who has attempted to replicate Professor Weber's discovery has failed.

Now there is no suggestion that Joseph Weber reported bogus results; his integrity is beyond question. Nevertheless, the verdict of the scientific community is that gravity waves probably have *not* yet been detected. However great the authority of the researcher concerned, the world of science can never accept the findings of one person unless and until they have been duplicated by others.

It may take a very long time for such controversies to be resolved. A good illustration is that of the twenty-one years of research which culminated in the award of a Nobel Prize to Dr Roger Guillemin and Dr Andrew Schally in 1977. Their achievement was to demonstrate the existence of a series of hormones called releasing factors, produced by the brain, which regulate several different bodily functions. Such were the formidable technical difficulties in isolating these hormones that their elusiveness was even compared at one stage with that of the Abominable Snowman. Until other laboratories had confirmed Guillemin and Schally's results, the scientific community had little choice but to keep an open mind about their work.

Novice or Nobel laureate – it makes no difference. A scientist's work is credible only to the extent that it can be duplicated by others. Reproducibility is the key.

5 Controls

'Shall I compare thee to a summer's day? . . .'
Shakespeare, Sonnet 18

We have discussed the principle that scientific work must be reproducible in order to be validated. Experiments must also be *controlled*, in the scientific sense, in order to see if the results are due to the factor under investigation or to some property of the experiment. If clothes washed in improved powder A come out perfectly clean, it may be due to the new powder formula or it may be because the water was very hot or the machine was extremely efficient. In order to determine whether the result was really because of powder A, a batch of clothes should be washed under identical conditions in powder B, and in powder A in cold water by hand. The result of the three washes will indicate whether the best results are due to powder A or to hot water.

A few years ago London's evening newspaper, *The Standard*, decided to find out whether a new type of slimming aid really worked. So-called 'starch blockers' had been receiving enormous publicity because of their alleged ability to interfere with an enzyme in our body which digests starch. In theory, obese people keen on losing weight could take these new pills; continue to gorge themselves with starchy food like potatoes, bread and cakes; and yet shed pounds of surplus flesh. But while some authorities praised the starch blockers as highly effective, others were totally unconvinced. Who was right?

The Standard announced what it called a trial, designed to settle the matter. Six volunteers were selected and weighed. The idea was that they should take the pills for four weeks, and the investigators would see what their weight was afterwards. Following the month's trial, the human guinea-pigs were weighed again, and the newspaper published its findings and conclusions.

The outcome was as follows. One volunteer, a lady dancer, was 'absolutely delighted' with the starch blockers. Her weight fell from 9st 4lbs to 8st 10lbs during the trial and *The Standard* described her as a satisfied customer. Another, older, lady lost only 1½lbs and announced: 'For me the starch blockers were completely useless.' A third volunteer shed 5lbs of excess weight. She was 'quite pleased' with what she had achieved but did not attribute her successful slimming to the tablets. The fourth guinea-pig – the only man to complete the course – did not lose weight and was extremely disappointed. The other two members of the group dropped out of the project before it was completed.

So what should we conclude? Perhaps starch blockers are no good after all, as only one individual out of the initial six discarded a substantial amount of fat? Perhaps the tablets work for some people and not for others? Perhaps *The Standard* itself was right to concentrate on the delighted dancer. She did really well, and two of the other three subjects who kept on taking the tablets for the full month did shed weight too.

The dismal truth, in fact, is that nothing whatever can be deduced from this so-called trial. If we look more closely at *The Standard*'s investigations we find that it was fatally flawed. From the very beginning it lacked the elementary principle – that of the control – which could have made its results genuinely meaningful.

A control is, quite simply, a means of comparison. If you want to know whether a new type of lawn fertiliser improves the quality of your grass, you need to set up an experiment in which you apply the product to one area and leave another untreated. The second area, which must be identical in every way with the first except for its lack of fertiliser, is called the control. By comparing it with the treated lawn at the end of the experiment, you can be reasonably certain that any difference really is attributable to the new fertiliser you applied.

This element of control was totally lacking in the slimming trial. There was no standard of comparison. Nor was there any attempt to standardise all of the other factors which may have affected the outcome. A simple before-and-after comparison, therefore, could be highly misleading – a point well illustrated in the case of the young dancer who actually did lose some unwanted flesh. We do not know whether her weight was stable initially or not. Also it turns out that 'she ate absolutely normally, but cut out alcohol'. So there are at least two plausible alternative explanations for her success. First, her weight may *already* have been falling. In that case the period of the trial – with or without starch blockers, whether or not they had any real effect – would be certain to show the same result. Second, her renunciation of alcohol could well have accounted for the loss in weight. Some people gain a substantial proportion of their daily calorie intake from such drinks.

It is not easy, of course, to design a study of this sort with an utterly foolproof control: you could not compare an individual taking starch blockers with one and the same individual not taking starch blockers, over the same period of time and with everything else (diet, exercise, etc.) being equal. Such an experiment is impossible – at least until the day of human clones!

However *The Standard*'s unscientific investigators could have got much closer to the principle of controls, so that they could have drawn meaningful conclusions. They should certainly have checked that their volunteers' weights were stable initially and ensured that they did not alter their diet or behaviour in any way. Then they could have extended the trial, monitoring their guinea-pigs' weight during several different months when they were (a) taking the tablets and (b) not taking them. Better still, they might have compared volunteers during months when they were receiving either starch blockers or inactive 'placebo' tablets without being told which was which.

Best of all would have been a much more extensive study, using starch blockers and placebos, in which many individuals, matched for age, weight, diet and all other relevant factors, could have been cross-compared. The use of large numbers helps to 'iron out' the biological variation between one person and another which can confuse and invalidate a simple comparison between the two. Statistical comparison of the two groups would then have revealed whether there was any genuine significance between their respective weight changes.

This is how scientists, rather than newspaper editors, would have set about the task. Indeed, this is precisely the way in which all modern life-saving drugs, for example, have been tried and tested. If *The Standard*'s approach to medical science were universal, we would never have been able to prove that an antibiotic like penicillin really works. We would still be living in the Dark Ages.

6 Science and certainty

'Doubt is not a very agreeable status but certainty is a ridiculous one.'   **Voltaire**

As mentioned earlier irritation is caused by scientists speaking in public who will not give straight answers to what seem to be straight questions. One reason for such 'hedging', we learned, was that discoveries can never be truly accepted until they have been repeated independently. However there is another, equally vital, reason for avoiding black-and-white replies: in many cases categorical answers are impossible because of the nature of the evidence. Whether we are talking about the results of an experiment in high-energy physics, or the likelihood of a patient suffering side-effects from a new drug, the best we may be able to provide is an estimate of *probability* not an assurance of *certainty*.

Consider the health hazards of smoking. These are now well-established and, in one sense, beyond question. Doctors and health educators are amply justified in trying to persuade people either not to start smoking or to give up the habit, if they want to keep fit and avoid certain diseases. But the dangers can be expressed only in probabilities, not in absolute terms. For example, smokers run about double the risk of contracting heart disease as compared with non-smokers; they are several times as likely to develop chronic bronchitis, and some 25 times as likely to contract lung cancer.

Given two populations of people, therefore, one consisting of smokers and the other of non-smokers – but identical in all other respects – one can make certain firm predictions. For example, there will be about 25 times as many lung cancer victims in the first group. What we can *not* do is to forecast that figure with total precision. It might turn out to be 26 times, or 24 times, or even more or less. Nor can we make any firm prophecy whatever about a single individual. There will always be fortunate folk who, against the odds, smoke 40 cigarettes per day without suffering any obvious harm. And there will be occasional cases of lung cancer among individuals who have never smoked. In what the biologist Garret Hardin calls 'fate's lottery', the only dependable guide comes from the odds: an estimate of the chances, not certainties, of a particular outcome.

It is like tossing a coin. The chances of turning up heads or tails are 50–50. But if we throw 10 times in succession, we may get six heads and four tails, or even seven heads and three tails. Toss the coin 100 times and the figures will approximate more nearly to evens – say 47–53. With a thousand throws, the outcome will be even closer, perhaps 482–518.

What, then, would be the odds against turning up another tail if, out of ten throws, we have *already* had nine tails (assuming, of course, that the coin is perfectly ordinary and not weighted in any way)? Would the chances of completing an amazing run of 10 out of 10 tails be nine to one against? Nine to one on? Ten to one against?

The answer, in fact, is evens. Whatever has happened to the first nine throws, the chances of getting a tail or head with that last toss of the coin remain at 50–50. Likewise with those medical examples, where the odds definitely *are* weighted: we can make no cast-iron prediction about what is going to happen to a single individual.

This explains two things about science. First, it shows why scientists have to collect such vast amounts of data. If scientists want to survey sulphur dioxide pollution in the air of two cities, they will carry out hundreds of thousands of identical tests in each place before taking the two averages from the two series and comparing those figures. Single analyses could be highly misleading. Second, it shows why experts invariably resort to probabilities when consulted about weather patterns, the spread of influenza epidemics, or earthquake activity in California. In such fields *estimates* of risk can be so impressive that they allow people to take evasive or protective action. Nevertheless, they remain estimates, based on the laws of chance. And they will sometimes prove to be hopelessly wrong.

The recognition of capriciousness and uncertainty at the very core of our understanding of the universe is what marks out modern science from that of the nineteenth century. A hundred years ago, the notion of cause-and-effect reigned supreme; derived from Isaac Newton and his contemporaries over a century earlier, it had become the bedrock of scientific method. Newton's most spectacular achievement was to have perceived that the planets were kept in their orbits by a sort of invisible celestial elastic: the force of gravitation. Extend that idea of causality into other areas, from geology to economics, and you soon have a universe which ticks over on the basis of endless sequences of causes and ensuing effects. This was the thoroughly deterministic model of nineteenth-century science.

Then came Albert Einstein, Max Planck and the other revolutionaries of the early-twentieth century. During the 1920s, they realised that the behaviour of sub-atomic matter, in particular the electron, could not be accounted for on the basis of a simple causal model. In 1927, Werner Heisenberg crystallised a totally new view with his 'principle of uncertainty'. He showed that in every description of nature there was some irremovable, essential uncertainty. The more accurately we measure the position of an electron, for example, the less sure we can be about its speed, and vice versa. So, as we cannot be certain about such a particle *now*, we cannot predict its future. At this point, cause-and-effect go out of the window.

Of course, Heisenberg's principle refers to very small particles and events but that does not mean that it is limited to the sub-atomic world. These are just the sorts of events which occur in our nerves and brain, for example, and in the macro-molecules that determine inherited characteristics. Uncertainty rather than Newtonian causality is the cardinal principle of our universe and statistics – the management and prediction of probabilities – is the tool we have developed to help us cope with that situation.

It certainly can be frustrating for a tidy-minded person, enquiring about some environmental hazard, to be answered only in terms of relative risk. Indeed, matters are even worse than this, because numerical estimates (like a 0.4 per million rate for death by tornado), are valid only if accompanied by further figures giving what are called 'tolerances' or 'confidence limits' which indicate the range of uncertainty attached to the estimate. If that sounds like scientists hedging their bets to an absurd degree, it is worth remembering one thing. Only by conducting experimental science on such probabilistic foundations have we been able to send men to the moon, obliterate smallpox, and encircle the planet with modern mass communications. No mean achievements.

7 Statistics

**'There are three kinds of lies: lies, damned lies
and statistics'
attrib. to Mark Twain,** *Autobiography*

'Statistics can prove anything', we are often told. But can they? While statistical analysis is an essential tool in most areas of research, it is also a weapon in the hands of lobby groups and politicians who want to persuade us of the truth of their case. Let us try to understand the difference.

'Despite the discovery of insulin, just as many people die of diabetes today as before.' That statement, which can be supported by statistics from every country in Europe, is literally true. Frederick Banting and Charles Best's wonderful work in isolating the sugar-regulating hormone, insulin, has had virtually no effect on the number of deaths from diabetes.

This is the type of ammunition often used by people who wish to decry the achievements of medical science and their arguments are usually based on incontrovertible data. So what is the snag?

In this case, there are two catches. The same number of people do now die from diabetes as before Banting and Best's breakthrough in 1922. However they now die beyond the age of 70, whereas previously they would not have survived their youth. Moreover, the world's population has doubled, from 2000 million to 4000 million, over the past half-century. These two facts are enough to show that the original statement, far from being straightforward, was wildly misleading. First, given the growth in population, it is quite wrong to quote *numbers* of individuals succumbing to a particular disease, as though they are strictly comparable; the only legitimate comparison is of *percentages* of people in the two groups. Second, to quote a death rate unrelated to age is indefensible; here it disguises the tremendous benefits that have unquestionably accrued from the introduction of insulin.

Despite their dubious reputation, statistical methods are essential in science. For example, if you want to survey changes in the plumage of wild ducks of a particular

species at one nesting site over several years, you will need to do a lot of counting. In order to simplify the task, you can use statistical techniques which allow you to count a sample rather than every single bird (the sample will have to be sufficiently large for valid conclusions to be drawn about the entire population). Another set of techniques will make it possible to decide whether any changes you notice from one year to another really are significant. The whole exercise is designed (a) to make manageable a complex, numerically large area of study and (b) to generate reliable information from the whole by examining only a part.

The difference between a scientist and a political lobbyist using statistics is one of motive. The scientist should begin by taking the widest possible view of a problem before focusing on a representative set of observations, experiments or measurements. The lobbyist always thinks of selecting data to buttress his case. A splendid example concerns the impact of immunisation against diphtheria, which spokesmen for the drug industry are fond of highlighting as one of their most astonishing success stories. At first sight such buoyancy seems amply justified. When diphtheria vaccination was introduced into Britain on a wide scale around 1940 the disease declined dramatically; the number of cases fell by 50–60,000 *every year* until about 1955, since when there have been only sporadic outbreaks. A killer disease has almost been eradicated.

This picture is woefully misleading. If we take a longer timescale and alter the criteria, the evidence looks very different. Diphtheria *deaths* actually went down continuously from 1860, when they numbered 1300 per year, to 1940, when the figure was under 300 per year. The steepest decline of all occurred between 1865 and 1875 – long before the diphtheria bacillus had been isolated and long, long before the development of immunisation.

A graph of the same period covering the commonest infectious diseases of childhood (for example, measles, whooping cough and scarlet fever), shows the death rate falling continuously. There is only a very small dip at the tail end reflecting the introduction of vaccines and antibiotics. So, while in no way disregarding the achievements of medical science, we do need to put them in perspective. Historically, far larger benefits have stemmed from greater prosperity which has led to improved nutrition, sanitation and hygiene. The relevance of this to health expenditure in Third World countries today should be obvious.

Statistical analysis as used by scientists provides genuine illumination, even though its objective is simply to get at 'the facts'. Statistics brandished by lobbyists, on the other hand, are employed in a determined effort to sway opinion, and often obscure or distort the truth. There was a good example of the latter case in an article published by the popular science magazine *Omni* (February 1980), which stated that 'the side effects of prescription drugs now equal breast cancer as a leading cause of death in the United States'. The comment came from a protagonist for so-called orthomolecular healing, as an alternative to conventional medicine. Again, as with the statement about diabetes deaths, this one seems to be crystal clear – an indictment of today's doctors and today's pharmaceutical industry.

However, like the diabetes sentence, it is tendentious because of the information *omitted*. The truth is that adverse reactions to drugs can be what are called terminal factors (and thus be mentioned on death certificates) in cancer, heart failure or indeed any fatal illness. There are two reasons for this. Firstly, promising but unproven treatments are undoubtedly tried out more commonly among terminally-ill patients (with their consent of course). If someone seems certain to die from a tumour which has not responded to established remedies, it is considered legitimate to offer new and essentially experimental drugs as a last hope. Secondly, even proven medicines, if they have real merits, also produce ill effects in a minority of patients and such effects are much more likely to occur in a person who is dying and seriously debilitated. So while it is strictly accurate to quote drug reactions as causes of death in cases of this sort, to do so out of context is a misrepresentation.

So, can statistics prove 'anything'? No, but they can be powerfully persuasive. To avoid being caught out let us learn a few simple principles from the examples given in this chapter. First, always bear in mind the motives of any individual or organisation issuing statistical data. Then ask yourself 'Have they selected these figures to convince me about something?' Consider the time scale and whether a longer one (going back earlier), would alter the picture. Try hard to imagine inferences other than the obvious one. See whether using numbers in place of percentages, or vice versa, would make any difference to the conclusions. Do not just accept figures you've been given: think about information that is missing and which could be relevant. If you take these precautions, the chances are that you will often see more than was ever intended.

8 Science and reality

'When I use a word, it means just what I choose it to mean.'   Humpty Dumpty in Lewis Carroll's Through the Looking Glass

So far in this book we have been mainly concerned with the process of science – its techniques. We have tried to show how scientists look at the world and how they tackle problems and add to the total knowledge of science. However, we should also examine the effectiveness of science as a means of helping us comprehend our universe and our place in it. That science does help us to understand the world about us is beyond question but it does so in a very special way – science *describes* our world in a detailed, specific manner, and rarely attempts to *explain* it directly in a way that would satisfy a philosopher looking for absolute answers. We may search for truth but science gives us knowledge instead.

A typical illustration is given by Newton's Laws of Motion. In developing these laws, Newton used the concepts of gravity and force-at-a-distance to describe the motion of the planets; the concept of gravity (defined by Newton as the force of attraction between two masses, which is proportional to the product of those masses divided by the square of the distance between them) does *describe* and predict the motion of the planets with remarkable accuracy but it does not *explain* such motions. The concept of gravitational attraction as an 'invisible rope' holding the planets in their orbits is no more satisfying as an explanation of their behaviour than the concept of the planets being propelled in their paths by invisible gods. Newton himself was unhappy with the concept of 'force-at-a-distance', perhaps associating it with astrological beliefs in the mystical influence of the planets. In a letter to Richard Bentley, another scientist, he wrote: 'That gravity should be innate, inherent and essential to matter, so that one body may act on another at a distance . . . is to me so great an absurdity that I believe no man, who has in philosophical matters a competent faculty of thinking, can ever fall into it' (25 February 1693). Indeed one of Newton's

24

contemporaries, Bishop Berkeley, referred to the concept of force as 'an occult quality'.

The word 'gravity' describes a mathematical idea whose sole function is to *describe* the behaviour of the planets. It explains nothing. Bishop Berkeley pointed this out at the time (see Fig. 1). He went on to claim that in physics there is no 'causal explanation' of reality based on some 'hidden nature' of things because such 'hidden or true natures' do not exist; the appearance of things *is* their reality. According to this view the role of science would be to establish certain universal principles – the 'laws of nature', which would describe the behaviour of things but would not attempt to explain them in a manner satisfying to a philosopher. The question of whether they were true or not would not arise, because such hypotheses could be

George Berkeley (1685–1753) . . . maintained that material objects only exist through being perceived. To the objection that, in that case, a tree, for instance, would cease to exist if no one was looking at it, he replied that God always perceives everything; if there were no God, what we take to be material objects would have a jerky life, suddenly leaping into being when we look at them; but as it is, owing to God's perceptions, trees and rocks and stones have an existence as continuous as common sense supposes. This is, in his opinion, a weighty argument for the existence of God. A limerick by Ronald Knox, with a reply, sets forth Berkeley's theory of material objects:

> There was a young man who said, 'God
> Must think it exceedingly odd
>> If he finds that this tree
>> Continues to be
> When there's no one about in the Quad.'

REPLY

> Dear Sir:
>> Your astonishment's odd:
> *I* am always about in the Quad.
>> And that's why the tree
>> Will continue to be,
> Since observed by
>> *Yours faithfully,*
>>> GOD.

Bertrand Russell, *History of Western Philosophy*

Figure 1

regarded as fictions – mathematical models (see Chapter 3) – that do not really exist (see Fig. 2).

Looked at from this viewpoint, one could have a pair of contradictory hypotheses which both describe the same phenomenon, and provided they both give valid results in describing the behaviour of something, then there can be no grounds for choosing between them. For example, we live quite happily with the two hypotheses concerning the nature of light, one of which assumes that light is a particle, the other that it is a wave: Sir William Bragg said, 'We use the classical laws on Mondays, Wednesdays and Fridays, and the quantum laws on Tuesdays, Thursdays, and Saturdays.' Since both hypotheses can be demonstrated to describe the behaviour of light correctly under certain circumstances, both are valid, but which is *true*?

'. . . and yet, it moves'.

It was this difference of viewpoint – hypothesis versus reality – that was at the heart of the conflict between Galileo and the church: for while the church was prepared to accept the Copernican theory of a sun-centred solar system as a *mathematical assumption* which simplified calculations of planetary orbits, it could not stomach Galileo's continued insistence that *this is the way the solar system really is.* Copernicus settled for hypothesis; Galileo held out for reality, and suffered the consequences.

Now it is argued by some scientists that this whole conflict was a red herring, since Einstein has shown that all motion is relative, so that it is meaningless to argue about absolute motion: we can choose any place we wish to specify as being (relatively) at rest. Frankly, this argument is nonsense. Nobody really believes that the earth is at the centre of the solar system – even hypothetically. Why? Because science has given us a broader context within which to view the altercation: a universe in which there are multitudes of galaxies, with myriads of stars rotating about the hub of each galaxy, some of these stars having planets revolving around them, and with our own sun just one of these stars. Against this background of knowledge, the assertion that the earth is the immovable centre of the universe is obviously and quite simply *not true*. It does not accord with our observations. *This* was what Galileo understood when he said: '. . . and yet, it moves'.

Figure 2

Berkeley asserted that the question was irrelevant, and he was correct in that scientists do not waste time discussing the point. They use either or both hypotheses whenever it suits them. However, Berkeley was on less solid ground in his claim that science, since it 'only' describes the external appearance of things, could not concern itself with the 'true or real nature' of our world because, he asserted, there *was* nothing physical *behind* the surface appearance of things. Physical objects *are* what they appear to be. On these grounds, Berkeley rejected the early attempts to develop an atomic or corpuscular theory of matter, because they attempted to explain the 'world of appearances' by constructing an invisible world of 'inward essences' *behind* the world of appearances.

Now we know that, in describing the world about us, science, and particularly physics, *is* concerned with 'appearance' but not in the sense understood by Berkeley, that is, that which we observe with our limited and unaided senses. Technology has provided science with tools which extend our senses to a degree unimagined by Berkeley. Today, we view the world with infra-red, ultra-violet, radio and even X-ray 'eyes', and the appearance of our world viewed through these extended senses is every bit as real as its more limited appearances to our own eyes; because of these extensions to our senses we can observe physical realities, like atoms, behind the appearance of reality that we normally see. We can now obtain photographs of the larger atoms and molecules and they are as *real* as the surface texture that we can see and feel.

So science does explain the nature of our world in a manner which Berkeley insisted it could not do. Of course our explanations are still descriptive explanations as distinct from the absolute explanations sought by metaphysics and philosophy. They do not provide a complete answer to our question 'What is . . .?', but lead on to further, more searching questions:

What is matter? Matter is made up of molecules.

What are molecules? Molecules are made up of atoms.

What are atoms? Atoms consist of protons, neutrons and electrons.

What are protons, neutrons and electrons? They are constructed from quarks and leptons.

What are quarks and leptons? They are . . . and so on.

There is no sign that there is an ultimate end to such questions, only a seemingly endless series of questions and answers gradually converging on some sort of ultimate truth which will never be reached. As human beings of limited intelligence, surrounded by a cosmos of infinite complexity, can we expect more?

9 Theory and experiment in science

'Don't bite my finger – look where it's pointing.'   W.S. McCulloch

Which comes first in science, theory or experiment? This 'chicken and egg' conundrum has interested scientific philosophers for many years. It is not an idle question because by considering it we can reach important conclusions about the nature of science. Nevertheless, many practising scientists tend to dismiss such 'philosophical' questions as a waste of time because they are not answerable in a satisfactory manner. Scientists prefer to ask questions that *can* be answered.

Historically, scientific method started with simple observations. In ancient times, these observations were mainly concerned with the sun, moon and stars, for reasons associated with agriculture, navigation and religion. Such observations, by the time of the Babylonians, were very detailed and comprehensive. Along with this sophistication in astronomy went comparable skills in measurement and mathematics. Using this knowledge, the Greek philosophers, and in particular Plato, formulated theories: not of the way the world was but of the way it *ought* to be. The Greeks were rather high-minded about nature, they had very definite ideas about what an ideal universe should be like and it was up to the universe to match up to the ideas. When it did not, well, that just showed what in imperfect world they lived in. Such an attitude, by today's standards, would not be regarded as particularly scientific. Nevertheless, the detailed observations of nature, combined with the construction of theories explaining such observations (to some degree), represented a sort of passive proto-science which has its counterparts today (for example, statistics). The idea of conducting controlled experiments in order to verify their theories never really occurred to the Greeks, or if it did they considered it a rather unworthy activity for philosophers.

It was Friar Roger Bacon (1214–1294) who first introduced the idea that nature should be studied by means of controlled experiments. His concept of science was the modern one of devising experiments which, so to speak, back nature into a corner and force from her the answers demanded by the experimenter. This is not an easy thing to do because nature tends to be a hostile witness in such circumstances. It is no good putting her on the stand and saying 'Now tell us all about yourself'. Like most reluctant witnesses she has to be pinned down by questions which say, in effect: 'I put it to you that such and such is the case. Answer yes or no.' The technique of devising such experiments is an art in itself. It requires in the experimenter the gift of being able to ask the right questions and the quality of persistence in order to pursue the problem to the end.

In order to devise effective experiments of this type we have to have in our minds some sort of hypothesis at the outset. The purpose of the experiment is then to ask: 'Does this hypothesis accord with reality?' This is how modern science progresses, with hypotheses followed by experiments which are used to refine the hypotheses or to lead to new and better ones. In this sequence it is irrelevant to ask which comes first, hypothesis or experiment. They alternate in an infinite series that represents the feedback process that we call science.

It would be very satisfying to believe that all scientific knowledge had developed smoothly and progressively in this way. Unfortunately, the truth is more disconcerting and much messier. First, scientists may become so attached to a particular hypothesis that they reject the results of experiments which contradict it. They usually do this by categorising such results as 'anomalous'. Second, if we think of scientific progress as a road of discovery, along which scientists methodically plod towards the goal of ultimate truth, then we must also imagine that road as having innumerable forks, most of which lead to dead-ends or impenetrable thickets. Once one is actively engaged in research there is no way of knowing whether one is on a false trail or not. Third, some scientific discoveries have resulted from sheer luck. An example is recounted by J. D. Bernal in *The Extension of Man*: the Chinese of antiquity used to predict the future by means of a spinning spoon placed at the centre of a 'geomancer's' board (see Fig. 1). Various predictions about the future could be made, depending on which way the spoon finished up pointing. Now as it happened, some of these spoons were made from lodestone (magnetite), and the Chinese soon discovered that these spoons always finished up pointing north-south. Thus the compass was discovered.

Finally, it appears that some of science's most famous

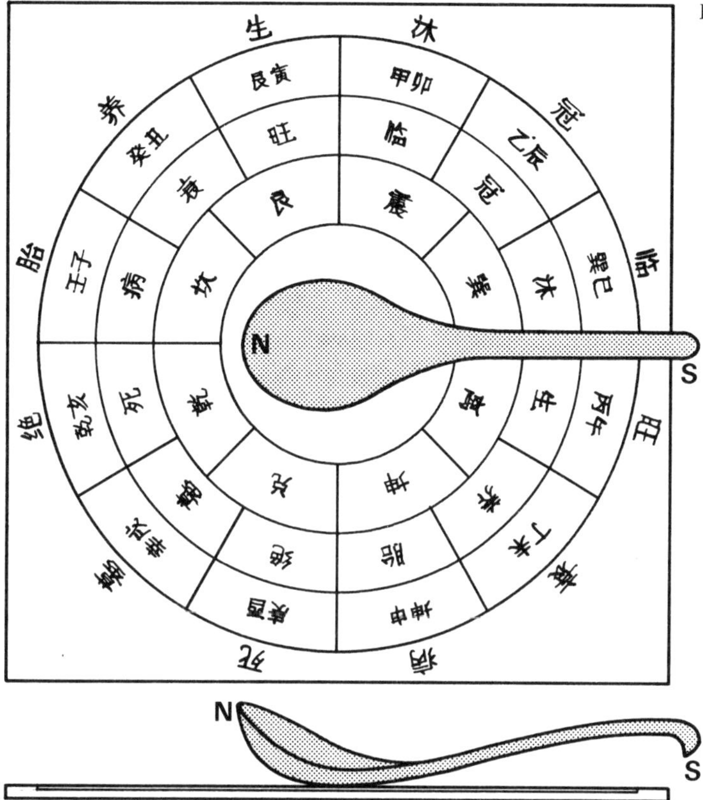

Figure 1

A reconstruction of a geomancer's board – the earliest form of compass (Han period, AD 100). The spoon was cut from lodestone.

Figure 2

insights were derived from neither experiment nor observation. In fact, some of them resulted from basically unscientific ideas that appear to us as strange or even dotty. For example, in the seventeenth century Copernicus' idea of placing the sun rather than the earth at the centre of the universe was a result not of new observations but of a semi-religious belief, inherited from Plato, that that was the right place for it, since it represented the greatest worth of all things in nature and should therefore be at the centre of things. Then again, Kepler (1571–1630) a brilliant mathematician and astronomer, was steeped in astrological lore and obsessed with uncovering the 'harmony of the spheres'. Kepler's famous laws were to a great extent inspired by his belief in a power or 'influence' which emanated from the sun and steered the planets, including our own earth: a belief which is fundamental to astrology. It seems that many of our scientific idols have feet, if not of clay, then well-muddied with mysticism and folklore; even the great Newton cannot be excluded, having spent a lot of his time on the study of what we would now call 'alchemy'. In more recent times, scientific views have been affected by political and moral beliefs. The study of genetics in the USSR was side-tracked for several decades after 1948, when the ruling Communist Party, prompted by the biologist Lysenko, backed the teachings of Michurin and condemned the work of Mendel. Michurin's work was held to be more ideologically correct and this fact was allowed to over-ride scientific truth. Even today scientific argument about such matters as inherited intelligence and race are coloured by deeply-held beliefs that have nothing to do with science.

You can see that our original question about theory and experiment in science has led us into uncovering aspects of science that are rather different from those normally portrayed, which is what was intended. Science, like most human activities, is a rather messy business in practice.

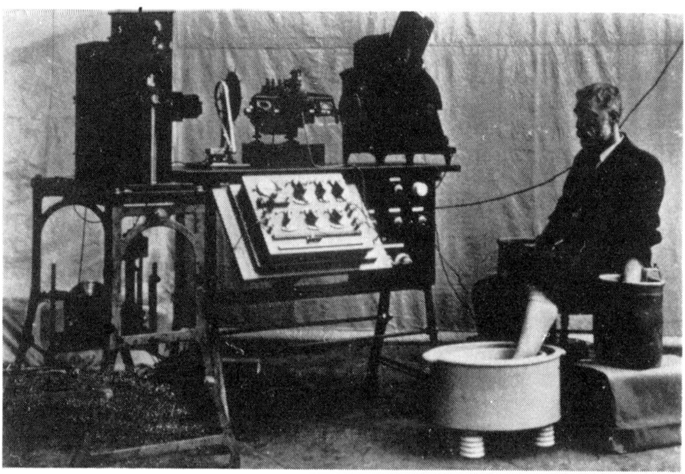

In real life, experiments come in a variety of types and are used for a range of purposes. In thinking about scientific activities we can define a list of categories which encompass most of the things that scientists do. We can label them:

Observation, cogitation, exploration and investigation.

In this list only the last two categories are what we would normally regard as 'experiments'.

Observation is, in fields such as astronomy, the only form of activity that is permitted, since we have no means of controlling the objects of our interest – the planets and stars. All we can do is form hypotheses about what makes the universe behave as it does and then see if the facts as we observe them fit the hypotheses. We are, however, permitted the luxury of *cogitation* along the lines of 'let us imagine what would happen if . . .', and as we shall see in the chapter on 'thought experiments', this type of thought process can be a powerful tool. Probably the greatest scientific theory outside astronomy based almost exclusively on observation must be Darwin's theory of evolution.

Next we have *exploration*, which is that type of limited experiment which we can conduct on objects with which we have contact but only limited control. Such experiments in their most primitive form are explorations in the literal sense of voyages of discovery, and are characterised by the fact that in most cases one has no clear idea of where one is going. The child with a new chemistry set is conducting this type of experiment when he or she wonders: what would happen if I mixed *this* powder with *this* liquid? A good example of exploratory experiments is given by the early work of anatomists attempting, by dissection, to comprehend the working mechanisms of living things.

Finally, there is *investigation*: an experiment conducted for a specific purpose under conditions in which we have good control over the objects being studied. A typical example would be the discovery of Boyle's Law. The subject is a gas (any gas), and the variables are the volume, temperature, pressure and quantity of the gas. By devising ingenious laboratory equipment it is possible to hold some variables constant (in this case the quantity of gas and its temperature) and investigate the relationship of the remaining variables (the volume and pressure). The great majority of scientific experiments are of this type, although the more deeply we delve into nature the more difficult (and expensive) such experiments become.

10 Thought experiments

'Why sometimes I've believed as many as six impossible things before breakfast.'
Lewis Carroll, Through the Looking Glass

The phrase 'thought experiment' was coined by Einstein, who used the technique while working on his theory of relativity. Since this involved situations where objects move at speeds comparable to that of light, it was obviously not open to practical experiment. Einstein therefore devised experiments in his head which were simple enough for reliable conclusions to be drawn in the absence of experimental data. A simple example is sufficient to illustrate the technique.

Most people know of Galileo's famous experiment (which historians do not believe he actually performed), to demonstrate that objects always fall towards earth at the same rate, regardless of their mass (see Fig. 2). Prior to

Figure 1　Einstein

The initially surprising fact that all bodies in a gravitational field fall with the same acceleration (the inertia of the heavier of two bodies *exactly* compensating for the additional pull of gravity) was endowed with a deep physical meaning by Einstein, who in 1907 was trying to modify his special theory to accommodate the behaviour of accelerating objects. He imagined an observer in 'free fall' who had the presence of mind to conduct experiments. If such a single-minded observer released any objects they would remain *relative to him* in a state of rest or uniform motion in a straight line (neglecting air resistance). For such an observer the gravitional field would not exist, and any conceivable experiment conducted by him would justify his assuming that he was, in fact, at rest. It was this realisation, according to Einstein, which led him eventually to the general theory of relativity.

Galileo everyone assumed that heavier objects fell faster than light ones, it was commonsense. Galileo did finally convince people that bodies of differing mass do indeed fall at the same rate but if you are not inclined to take Galileo's word for it, and do not have a suitably tall building to hand, here is a simple 'thought experiment' that should convince you.

1 Imagine two spheres, one made of iron, the other of wood. Though they are the same size, the iron ball will be heavier – let us say four times heavier than the wooden ball. We wish to prove that, in spite of the weight difference, the balls will fall at the same rate.

2 First, imagine that we turn the woóden ball into an iron ball *of the same mass*. Since the mass is unchanged, we can be confident that this change will not affect the rate at which the ball falls. (The change of *size* will not matter because we can conveniently ignore the effects of air resistance in a thought experiment).

3 For the next step, we imagine the large iron ball being hammered into a solid lump in the shape of four small balls, each of which will be the same size and mass as the iron ball which we substituted for the wooden ball. All we have done here is change the *shape* of the heavier ball so this step will also not affect the rate at which it falls.

4 Now we mentally separate the four joined balls from each other and leave them connected only by loose string. The balls are still connected to one another and can therefore still be considered as one object. All we have done is change the nature of the connections, which will not affect the rate of fall.

5 For the last step, we can say that since our four balls are connected only by loose string, they cannot be interacting with one another in any significant way, so we may as well remove the string.

6 We now have five balls, all of the same size and mass, which must all, therefore, fall at the same rate. To arrive at this situation from the original state of affairs we have done nothing that would conceivably affect the rate at which the objects fall, so we can conclude, with a degree of confidence, that the original two balls of iron and wood will fall together.

In arriving at this conclusion we did not have to climb a single tall building.

There are many things in the world around us which we assume to be true without ever really thinking about them. In such cases, commonsense can result in misleading or false beliefs. Thought experiments are not only a useful

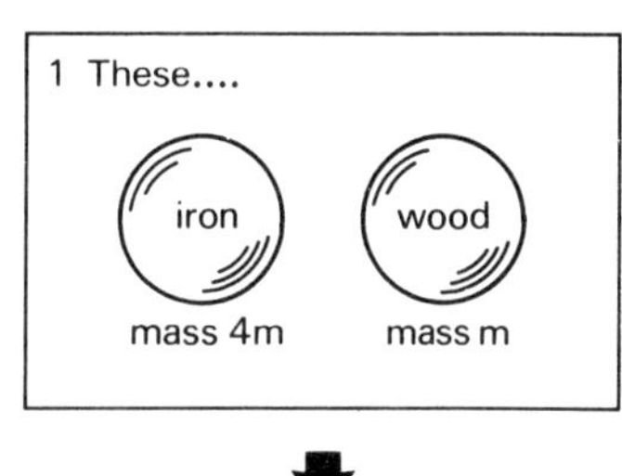

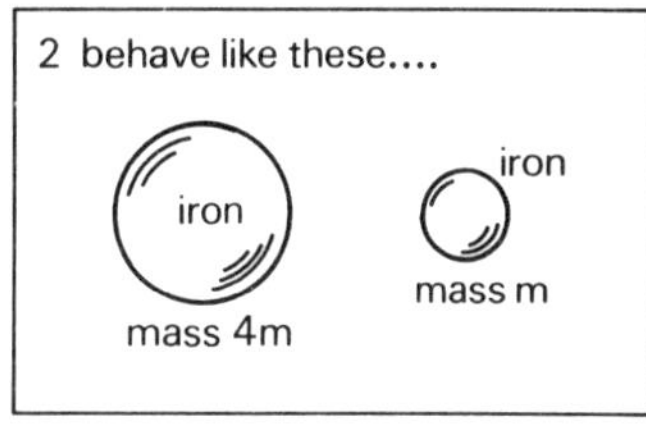

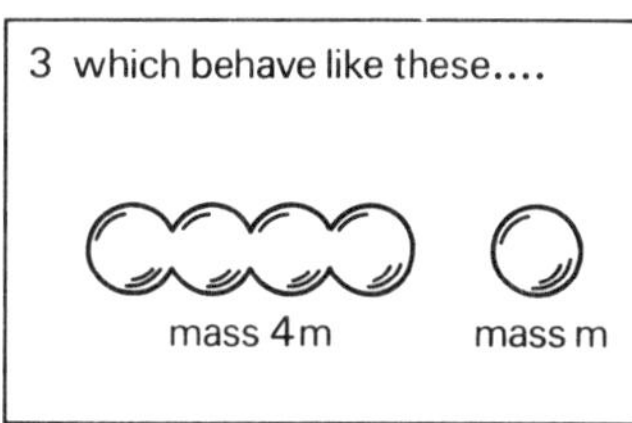

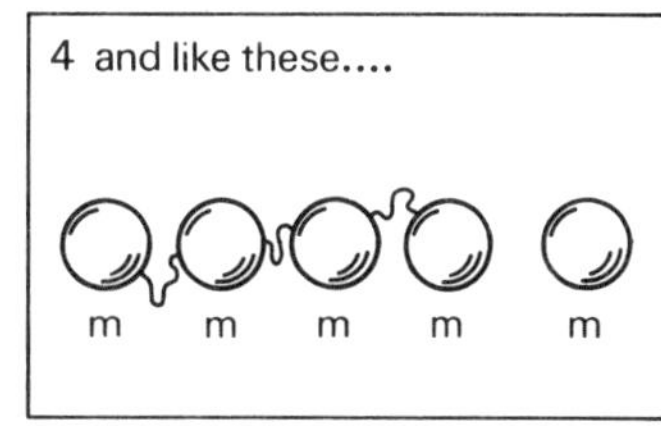

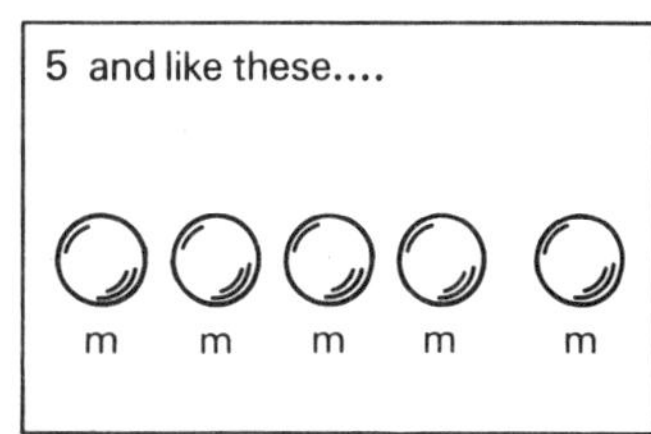

Figure 2

way of filtering out such misconceptions but often, as is the case in many areas of astronomy and cosmology, they are the only type of experiment that can be carried out.

11 Science and imagination

'. . . what do you mean, 'Eureka'?'   *anon.*

Most people, if they have pondered the question at all, would say that scientists are generally rational and artists are generally intuitive. This certainly accords with the stereotypes discussed in Chapter 1, but are there any scientific grounds for splitting human behaviour into two such groups? Observations on victims of brain damage indicate that there may be. For example, damage to the temporal or parietal lobes of the left hemisphere of the neocortex frequently impairs the ability to read, write, speak and do arithmetic. Comparable damage to the right hemisphere can impair three-dimensional vision, musical ability and pattern recognition. This suggests that those activities we call rational are associated with the brain's left hemisphere while those we consider to be intuitive or artistic have their source in the right hemisphere (see Fig. 1).

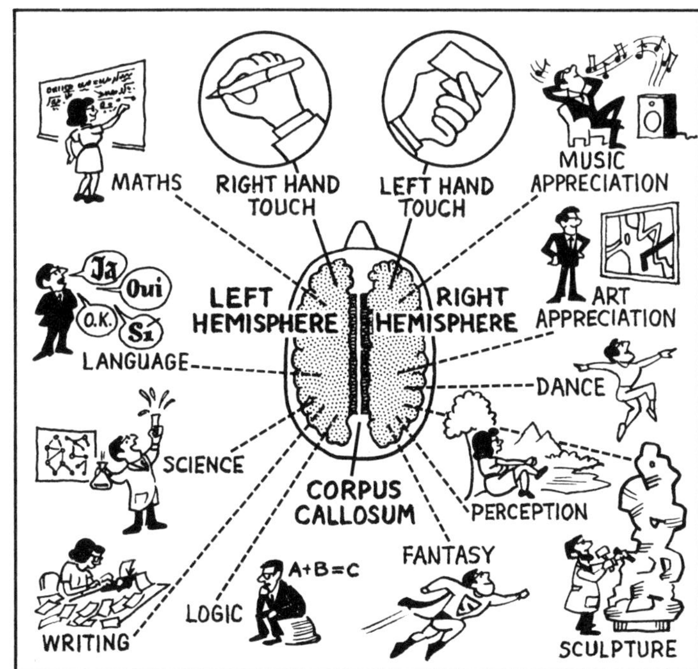

Figure 1

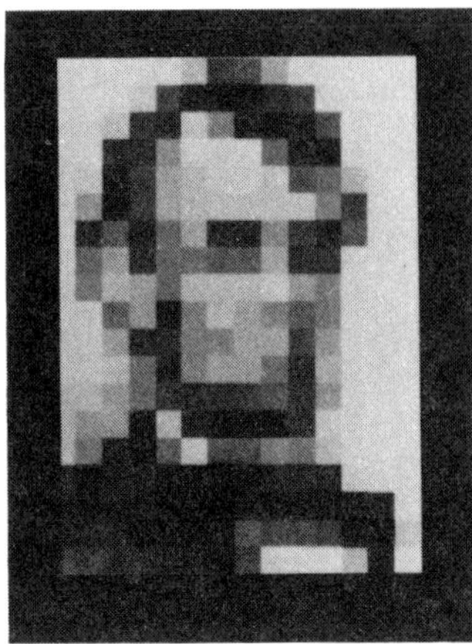

Figure 2 Pattern recognition. What does this pattern represent?

While such evidence is far from conclusive, it gives us at least some basis for splitting human behaviour into two types. So let us assume the validity of the two groups and ask whether rational, linear, logical thinkers really do make the best scientists? The short answer is 'No'. As we saw in the chapters on modelling and theory and experiment in science (Chapters 3 and 7), a crucial part of a scientist's job is that of proposing hypotheses which can then be tested by experiments. Now this hypothesising is no more than making up stories about how the world *might* be and as such is a very creative, intuitive activity. It requires an ability to see form and pattern on the basis of inadequate data (see Fig. 2). On the other hand the experimental verification of the hypotheses is very much a logical activity. So the good scientist needs a peculiar combination of scepticism and imagination, of fact and fancy.

In training scientists it is fairly easy to foster a proper scepticism but how on earth does one go about inculcating or developing imagination, and anyway, what do we mean by it? Imagination takes many forms. First, there is intuition – the inspired guess or the ability to jump to the correct solution as if by instinct. Often such intuitive leaps are justified afterwards as a logical process of scientific deduction but the fact remains that at the time they do not occur that way, at least not consciously (see Fig. 3). Intuition is invaluable to the researcher ploughing through masses of data, particularly in subjects of great intrinsic complexity such as biology and medicine. Good intuitive scientists can often get a reputation for being 'lucky', although luck usually has little to do with it. One example is the extraordinarily successful anthropologist Richard (Lucky) Leakey of the National Museums of Kenya who, in 1976, discovered a nearly complete skull of *Homo erectus* in geological strata one and a half million years old. Such 'lucky' finds have characterised Leakey's career. At the end of 1983 it was announced that Leakey had discovered the bones of what could be the ancestor of all modern apes and man.

Then there is *creativity**, in the sense of inventiveness, a

Figure 3

---

*There is another meaning of 'creativity', that of the ability to produce a 'creation'. In this sense, 'Hamlet' is the creation of Shakespeare: it is uniquely his work because no-one else could possibly have conceived it. Such unique personal creations hardly ever exist in science. Even Newton's Laws are so-called because Newton first propounded them, not because nobody else could have done so. Art is very much a collection of personal creations whereas science has almost none, only a growing body of 'universal' knowledge.

talent perhaps more valued in engineers and designers but an invaluable skill in every field. The purest example of this particular talent is exhibited by inventors, where it often appears to be completely independent of education or training. Many inventors are rather odd, solitary characters, not well-suited to working in a team. A few notable scientists have been like this but their reputations as scientists have usually suffered as a consequence, an example is Nikola Tesla. Tesla was a talented inventor, he devised the induction motor, new forms of dynamos, transformers, condensers, arc and incandescent lamps, but in his later years he became more and more of a recluse and preferred to announce his inventions to journalists rather than fellow scientists. Naturally this did not endear him to the scientific community.

Inventiveness is a skill much prized in experimenters and technicians. An example of an exceptional technician is Robert Hooke, who was assistant to Robert Boyle (of Boyle's Law). Hooke was a marvellous experimenter and inventor and historians strongly suspect that he carried out most of Boyle's experiments. (It is significant that after they parted company Boyle did no more experiments.) The great Leonardo da Vinci was the ultimate technician. He did not make any great contributions to science but his abilities as an inventor were stunning.

Finally there is insight, probably the most valuable and rarest of the imaginative skills. Insight is concerned with depth and breadth of understanding and is closely associated with what most people would call wisdom. This is the attribute we recognise as common to the few outstandingly great scientific thinkers: Galileo, Newton, Darwin and Einstein. It is characterised by breadth of vision and, unlike either intuition or creativity, tends to spill over into abilities and skills outside the scientific sphere. A person can be intuitive and creative and yet have relatively narrow interests and a shallow understanding of the broad scope of science and human knowledge as a whole. Many quite successful scientists fall into this category. The few outstanding scientists who exhibit what we have chosen to call insight are worth listening to even when they are not talking about science. Two examples from our own generation are Jacob Bronowski and Peter Medawar. Bronowski is now well known to the general public for his excellent book and television series *The Ascent of Man*. Medawar is well known to scientists for his Nobel Prize-winning work in immunology and for his essays on science. His *Advice to a Young Scientist* is an essential read for any prospective scientist.

Of course, there are examples of eminent scientists who

combine some or all of these forms of imagination. Thomas Young was not only one of the founders of the properties of materials (Young's Modulus), he was also a pioneer investigator of the wave theory of light; made a significant contribution towards the translation of the Rosetta Stone (a tablet containing an inscription in two forms of Egyptian hieroglyphics and in Greek, which supplied the key to the ancient inscriptions of Egypt); and invented the ophthalmoscope (the instrument which opticians use to look into the eye). It is perhaps encouraging to note that, despite his brilliance as a scientist, Thomas Young was one of the worst scientific writers of all time. Anyone taking the trouble to seek out his original enunciation of the Modulus of Elasticity will discover that it defies all attempts at clarification.

So we see that in the practice of science imagination is an essential ingredient when combined with a proper degree of critical judgement and that without judgement, a scientist's theories can be wild and foolish; without imagination, however, his or her output may be barren. Now we all have a degree of imagination and the beauty of science as a profession is that it permits us to use what we have to the full. The vast majority of people entering science are not going to impress any significant footprints in the sands of time but they can have full and rewarding professional lives that tax their talents to the limit.

12 Science and pseudo-science

'What we need is not the will to believe, but the wish to find out.'   **Bertrand Russell**

How can we distinguish between science and pseudo-science? This is a vital question for scientists and those interested in science at a time when the traditional sciences are under increasing attack from scientologists, creationists and a whole assembly of new cults which ape the scientific method and use scientific jargon.

The following discussion examines the criteria which can be used to distinguish genuine science from its scientific-looking and -sounding imitators. In order to make the distinction we must first analyse what it is that makes 'the scientific method' special. If there is one thing that distinguishes science from all other human activities, it is the way in which scientists are prepared to change their minds. They may do so with the greatest reluctance but, when the weight of evidence piles up sufficiently, the scientist will give up his pet theory and accept a new idea. It is because of this attitude that science is one of the few professions in which there is no discredit in being wrong.

How did such a unique attitude come about? Human beings, including scientists, are notorious for the tenacity of their beliefs for which they are quite often prepared to kill or be killed. Only twice in human history did societies emerge which had the openness of mind and the self-confidence of spirit to provide a suitable seed bed for the growth of what we now call 'the scientific method'. The first time was in ancient Greece, which produced the miracle of Euclid's geometry – a triumph of logical thought but limited by its relative indifference to the real world (see Fig. 1). The second time was during the Italian Renaissance, when that supreme realist Galileo Galilei showed how all knowledge must grow out of experience:

> I think that in discussions of physical problems we ought to begin not from the authority of scriptural passages, but from sense-experiences and necessary

An interesting reason for the phenomenal success of the Greeks in creating the foundations of modern science is given by J. D. Bernal in his book *The Extension of Man*. Bernal suggests that, although the Greeks were inferior to the Babylonians in mathematics, having no proper number system (they used letters for numbers), they had a strong advantage in that they had no traditions, no respect for the gods of the various places which they visited during their travels:

> If you lived all your life in the city of Heliopolis, the city of the sun, you naturally had a very strong fixation on the local sun god, Ra; but if you only visited it occasionally and went on to other places, you built up a picture of the universe in which the gods were left out. If you had no gods and no proper tradition, you had to do some thinking for yourself and, essentially, that was what the Greeks did.

> The freedom of thought associated with the Italian renaissance was also, to some degree, a result of trading contacts, which familiarised renaissance man with the multiplicity of traditional beliefs held by other peoples around the world.

Figure 1

demonstrations . . . Nor is God any less excellently revealed in Nature's actions than in the sacred statements of the Bible.

Today, the scientific method is accepted as the principal means by which we attempt to understand our world. Perhaps the best way of thinking of the scientific method is as a *self-correcting process* which is used to extract reliable, predictive knowledge from the chaos of conflicting data and hypotheses that invariably exists at the frontiers of scientific research. This process of continuous change, in which ideas compete with one another for predominance, is only possible because all good scientific ideas and theories are vulnerable. In fact they are designed to be so. A good scientific theory has two principal characteristics: first, it explains the observed facts and does so in the simplest possible manner (this need to supply the simplest explanation consistent with the evidence is frequently referred to as Occam's Razor*); second, it must be testable

---

*Einstein put it best: '. . . the grand aim of science . . . is to cover the

– in other words, refutable. For example, compare Freudian psychiatry and Einstein's Theory of Relativity – both theories have broad powers to explain observed facts, the former of human behaviour, the second of physical behaviour. In both cases, this ability to explain a wide variety of facts is regarded as evidence of the truth of the theory. Is Freudian psychiatry, then, a scientific theory? It is not, for the simple reason that it is not *testable* in the way that Einstein's theory is. (Indeed, a major defect of modern psychiatric theory is that it has made itself almost completely invulnerable to any form of testing: so much so that, if you have doubts as to its validity, they may be explained as having been produced by defects in your own personality – such defects, of course, being treatable by psychoanalysis.) The Theory of Relativity, on the other hand, can and has been tested. As Einstein himself said: 'No amount of experimentation can prove me right, but one experiment can prove me wrong.'

So we have our first two criteria for identifying a good scientific theory:

(i) *It explains the facts in the simplest possible manner.*

(ii) *It is testable.*

Because of these characteristics, particularly the second one of testability, science changes continually. It is this changing character, this ability to accept and incorporate new ideas, that distinguishes science from pseudo-science. Think of how science has changed in the last thirty years, particularly in biology (the discovery of DNA), geology (tectonic plate theory), and cosmology (the resurgence of the big bang theory), and you begin to appreciate how rapidly scientific views of the world have changed. In the pseudo-sciences, on the other hand, viewpoints are maintained doggedly for their own sake. A typical example is astrology, which has changed little in the last 2000 years. This characteristic gives us another indicator to distinguish science from pseudo-science: *science is subject to change, psuedo-science is not.*

If you listen carefully to discussions concerning the merits of science and any of the pseudo-sciences you will soon detect a pattern emerging in the type of argument used to defend the pseudo-scientific belief, and this can give us further clues to add to our list of criteria. In order to appreciate these clues we must turn our attention from what is being said to the reasons for saying it, and this brings us to the different meanings of the word 'belief'.

In general, scientists are not prepared to give their

---

greatest possible number of empirical facts by logical deduction from the smallest number of hypotheses or axioms.'

42

unqualified belief to any theory, even when it has become so well established that it could with some confidence be called a scientific law. At best they are prepared to believe provisionally in the truth of a given theory; they will accept the current theory until a better one is propounded. The belief required of followers of any of the pseudo-sciences, however, is of a quite different kind, so let us, for convenience, distinguish it with a capital – Belief. This type of Belief is made either as a conscious act of faith (as with religious Beliefs) or as the result of a response to some deep psychological need; whatever the reason, Belief is absolute and unchangeable, even in the face of overwhelming contradicting evidence.

Since such Beliefs are absolute, their holders use the traditional methods of argument that are common to all belief systems that are based on a commitment to believe, as distinct from a wish to understand. The characteristics of such methods of argument are well understood and can be summarised as follows: First, objections to a given Belief are countered one by one, never allowing them to collect into a 'critical mass' of rebuttal. In order to do this, the Belief system will have a built-in circularity of argument: conflicting evidence can be interpreted by reference to previous explanations of evidence. If these explanations are questioned, reference is made to other data covered by the theory, and so on. It is rather like using a dictionary – if you doubt the meaning of a word, the dictionary will provide you with a meaning in terms of other words, whose meaning you will not be expected to question: and so on . . .

Second, a Belief system can always be expanded to cover almost any eventuality however contradictory the new event may appear to be at first. A good example of this is the way in which the Ptolemaic (earth-centred) Belief system attempted to explain the irregular movements of the planets by resorting to an 'epicyclic' (wheels rotating on wheels) explanation which was absurdly complex but did explain the observations *within* the established theory (see Fig. 2).

Third, the followers of any of the pseudo-science Belief systems usually regard those who disagree with them as 'non-Believers' whose motives are essentially hostile. Therefore, in a dispute, Believers may resort to personal attacks which question both the competence and the motives of their opponents. Of course such personal attacks are not unknown in scientific disputes but they are nearly always concerned with subjects that lie on the boundaries of science. (The Postscript to this chapter discusses the relationship between science and the

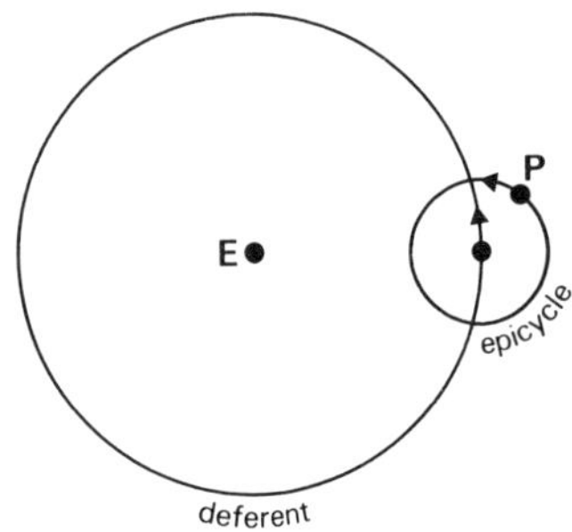

Figure 2   Diagram of an epicycle

quasi-sciences.) A good example is the seemingly-endless argument concerning the inheritability of intelligence. The fact that this subject impinges on ideological beliefs about racial differences and the equality of man ensures that the debate is frequently ill-tempered and personal. The very nature of this particular dispute assures us that we are not in the realms of mainstream science.

Finally, in our attempts at distinguishing between science and pseudo-science we must face the problem of having to make some sort of judgement about the people involved. Scientists are inevitably influenced by what their fellow scientists think, particularly those of some eminence. This does lead to a general conservatism in science since eminence tends to go with seniority but there is wisdom in this. Perfect objectivity is impossible, when evaluating data we invariably take into account the character of the source. Objectivity does not consist in giving equal weight to all statements, just as in a court of law the jury does not achieve justice by giving equal weight to the statements of both mugger and victim. An example is instructive: several years ago a group of scientists was formed to investigate the authenticity of that most famous of relics, the Turin Shroud. The project was called the Shroud of Turin Research Project (STURP). Recently a number of articles and books have been published by some STURP members claiming, or at least implying, that the shroud is genuine. Thus we have the astonishing example of a group of scientists apparently proclaiming the validity of a supernatural occurrence. Since science itself is an attempt to explain the natural world in its own terms, without resorting to supernatural explanations, this is a remarkable situation. In the absence of all the facts available to STURP members, what is an impartial scientist to make of these claims? His first instinct will be to check on the degree of impartiality of the group members. If he does this he will find that only one of the forty scientists associated with STURP is an agnostic. In fact, STURP is a volunteer organisation and it has attracted scientists who are predominantly religious believers and may not be predisposed to scepticism on this issue. This one fact alone will require the impartial scientist to regard the claims of such a group with profound suspicion, not because scientists with religious beliefs cannot be good scientists – that is manifestly not the case – but because of the predisposition to self-deception to which all committed groups, scientists or not, are vulnerable when faced with an issue about which they feel strongly.

In the search for a set of criteria to distinguish science from the pseudo-sciences we have considered the nature of 'the scientific method'; looked at some of the ways in which people come to believe things and how this shows itself in the types of argument that they use; and finally, with some reluctance, considered the necessity of weighing an argument against the calibre of the person using it. As a result, we arrive at the following set of criteria:

Principle characteristics of science
1 Change is a primary attribute of all science and, as a result, a scientist's belief in his view of the world is always tentative and provisional.
2 Scientific theories explain the facts in the simplest possible manner.
3 Scientific theories are testable.
4 Evidence for the refutation of scientific theories is actively sought.
5 Alternative theories are permitted a full hearing.

Principle characteristics of pseudo-science
1 Belief is absolute, unchangeable and held as an article of faith.
2 Beliefs tend to be idiosyncratic and unnecessarily convoluted.
3 The assertions and theories of pseudo-science are usually untestable.
4 Evidence contradicting the theories is either ignored or attempts are made to discredit the source.
5 Alternative beliefs are attacked by a series of argumentative devices that are clearly recognisable.

Of course, this list is not definitive – its edges are fuzzy. Respected scientists do, from time to time, behave like fairground charlatans and even, in rare but infamous cases (Piltdown Man, Sir Cyril Burt's identical twins data), cheat. Nevertheless, used sensibly, it is a fair guide, particularly when used in conjunction with an assessment of the character and motives of the participants in any dispute. Naturally one must apply commonsense in interpreting the criteria: for example, scientific ideas are open to change but there is a difference of degree between open-mindedness and gullibility. Commonsense has to be used in distinguishing sensible alternative theories from those of the 'flat earth' variety but even here commonsense can play us false. Until Galileo demonstrated otherwise it was 'commonsense' to assume that heavy objects fall faster than light ones (see Chapter 8). Another example: in 1772, at a meeting of the Académie des Sciences in Paris, Lavoisier, the father of modern chemistry, said (of meteorites), 'These rocks cannot have fallen from the sky because there are no rocks in the sky.'

Fortunately, in the genuine sciences, such misconceptions are amenable to change once sufficient evidence is available and scientists will adapt to new ideas, even when they dislike them intensely, because, in the end, good scientists have a greater commitment to science as a whole than to any particular concept contained within it.

POSTSCRIPT *Science and the quasi-sciences*

We have been concerned in this chapter with the task of distinguishing between science and the pseudo-sciences; that is to say, between science and those systems of belief which are, at heart, hostile to the spirit of the scientific method. There is, however, another distinction that it is important to make and that is between science and the neo- or quasi-sciences, by which we mean those disciplines which are committed to the principles and concepts of the scientific method, and attempt to apply them to fields of study where their suitability and effectiveness may be open to question. Such fields of study include sociology, behavioural psychology, educational psychology and economics.

It seems that all too often attempts to apply scientific methodology in these fields have an unfortunate effect: they tend to succeed as science in proportion to the degree to which they trivialise the subject to which they are applied. Unwarranted general conclusions drawn from such applications can inflict considerable damage on society as a whole. One example of this is the influence on our educational system of the naïve belief that intelligence can be measured on a simple linear scale. One can understand the attraction of simple numbers, particularly as they are useful in giving a sense of false security to those charged with making decisions. However, in those fields concerned with the difficult problem of measuring aspects of human behaviour, there is a tendency to pluck a representative number out of the melting pot of data and then kneel down and worship it. In educational psychology this number is the IQ.

Perhaps the most disturbing aspect of the formal enthronement of the IQ at the heart of educational policy in many countries is the way in which it is used to reinforce class preconceptions. Data which shows that the children of working-class parents usually score badly in IQ tests confirms the prejudices of those who believe in the rightness of class divisions and ensures that such children will be denied educational preferment. Thus the claim that children with high IQ scores invariably succeed in later life becomes a self-fulfilling prophecy. In fact the only incontrovertible statement that can be made concerning IQ scores is that they provide a measure of one's ability to take IQ tests.

13 Science and human values

*'Liberty means responsibility. That is why
most men dread it.'*
G. B. Shaw, Man and Superman

One of the most foolish criticisms made about science is that it is devoid of values. Because science is founded on objectivity – unlike the arts which are more personal, subjective pursuits – it is supposed to be amoral, a disinterested pursuit of truth. Once scientists don their white coats and enter the laboratory they are thought to renounce all human values and feelings. In the words of the post-war Labour politician Aneurin Bevan, they become 'desiccated calculating machines'.

There are two senses in which this is a damaging and inaccurate caricature. First, it suggests that scientific research is propelled forward by nothing more than its own internal logic – one discovery prompting the next logical set of possibilities. In fact, science is deeply affected by personal and social judgements. For example, every country has to work out how much money to spend on research, how much on the performing arts, and how much on the purchase of armaments. Such decisions reflect the values of particular societies at particular times. Similarly decisions within individual disciplines reflect the priorities and fashions of the group and its society, for example 'Should we allocate more of our medical resources to studies of mental illness, fundamental cell biology, or ways of combating the infirmities of old age?' The way we answer such questions depends upon our view of the relative importance of these problems, not just upon a technical assessment of the chances of solving them.

In recent years there has been a growing recognition that science is not value-free. Nor should scientific 'policy-making' be left wholly to scientists. Issues which formerly would have been left to scientists are now resolved in the political and social arena. As a result action on matters like the use of laboratory animals and the need to develop alternative testing techniques is now influenced much more directly by attitudes in the community at large.

However there is a second and more important way in which the conduct of science should not be portrayed as cold and devoid of human values. Jacob Bronowski highlighted it in his splendid book *The Common Sense of Science*, in which he demolished the view that science and morals are either opposed to each other or have no conceivable connection. Bronowski argued that, far from being amoral and ruthlessly logical, science actually generates ethical and social values. These values include an unusually acute regard for honesty, intellectual humility, respect for the revolutionary and the apparent crank, and stress on the importance of co-operation and social intercourse. Such values are not optional extras for working scientists: they arise inexorably out of the pursuit of science and the degree to which scientists live by them directly affects whether science flourishes or stagnates. For example, compare the attitude of scientists and politicans towards heterodoxy. Politicians have their principles (sometimes hardened into doctrine or dogma), and they measure new political movements against these, just as scientists assess new theories in relation to current orthodoxy. However, a politician can afford to denounce any opposition wholeheartedly and possibly make political capital out of the situation. But a scientist has to be much more careful. Scientific research is, by definition, a revolutionary activity and many of the major branches of today's physics, biology, chemistry and other disciplines began with claims or discoveries that seemed crazily out of line with contemporary opinion. So, without becoming soft-hearted or gullible, scientists should always take unconventional opinions seriously. Whilst being critical, they must pay respect and attention to rebels and revolutionaries – if only because so many of them have been right in the past.

Honesty is at a very high premium. Politicians may only tell the convenient facts – the end is supposed to justify the means; lawyers have to make the best possible case in spite of the facts or what lawyers call 'the merits'. Scientists have to be scrupulously honest at all times. As Bronowski stated 'For whatever else may be held against science, this cannot be denied, that it takes for ultimate judgement one criterion alone, that it shall be truthful.'

The reason is very simple. Science is a social rather than solitary pursuit (see Chapter 4), and people in one research establishment must be able to trust the information published by their peers elsewhere. A painter or sculptor in Washington, Cambridge or Melbourne may produce something that is disliked by artists on the other side of the world, but it does no harm. However if a researcher in America, Britain or Australia publishes fraudulent data in a learned journal, that act of dishonesty impairs the world's body of scientific knowledge. It introduces a serious flaw into the structure of science and may lead other scientists to waste valuable time and money in following up those bogus results.

As we discussed early in the book, research workers have not always lived up to the high standards expected of their profession. In their book *Betrayers of the Truth*, William Broad and Nicholas Wade gave examples of frauds committed by trusted investigators. In some cases scientists have consciously selected experimental results to make a point, conveniently neglecting data that did not fit their expectations. Some have gone further, inventing data from experiments which were never conducted at all. Even the work of Gregor Mendel, the pioneer of genetics, is not untainted. As R. A. Fisher showed in 1936, the Augustinian monk's tallies of plants in some of his classical pea-breeding tests were too good to be true. They must have been adjusted, probably by an over-zealous assistant, to conform with the expected ratios of short and tall plants, purple and white flowers, and other traits.

The fact that Mendel was undoubtedly correct in his basic theory does not excuse this departure from the scrupulous rules of conduct required by science. Indeed, the widespread anger that greeted the revelations in *Betrayers of the Truth* stemmed from scientists' feelings of disappointment that their craft should have been sullied with dishonesty. Whereas few of us would be surprised to read an account of chicanery in politics or double-dealing in financial circles, cheating and fraud in science are offences of the worst kind.

Finally, scientists demonstrate moral responsibility by choosing to draw the attention of their colleagues, and the public at large, to dangers arising out of frontier research. The most important instance in recent years was triggered off by the discovery in 1973 of simple test-tube procedures allowing microbiologists to combine DNA from quite different species of life, thus generating organisms that had never existed before. This development created two possibilities. One was that of harnessing the new science of 'genetic engineering' for beneficial purposes, for example, to produce insulin and other pharmaceuticals in bacteria, much more cheaply than was possible previously. The other was the spectre of horrendous organisms, inadvertently released, which might cause unstoppable epidemics and untreatable illnesses. It is to the credit of scientists led by Professor Paul Berg (who subsequently won a Nobel Prize for his researches in this field), that they decided to 'blow the whistle' and thus highlight these potential hazards. This precipitated government enquiries in several countries, and led eventually to the regulations which now specify the ultra-safe conditions under which such experiments are permitted.

14 Science and social responsibility

We have talked about the ideas and techniques of science and, in the previous chapter we touched on human values in science, but we have not really confronted the problems that may face scientists and engineers working in industry or for the government: problems connected with duty and social responsibility – to one's nation and to humanity. There is a good reason why we did not write on this subject, and that is because, as scientists, we dislike writing on subjects in which we have no particular expertise. (George Bernard Shaw once said that there's no fool like an educated fool, once he starts talking about subjects he has not been educated in.) In science, if you cannot do it, you do not pontificate about it – at least not if you want your comments to be taken seriously. In art, and particularly in literature, this is not the case. There is a branch of literature called 'literary criticism' in which books are written about other people's writings. However the critic is not expected to be either an expert in, or even a practitioner of, the form he or she is criticising. Now this is a perfectly valid attitude but it is one that makes scientists uncomfortable. In science, one is expected to have demonstrated a certain minimum competence *as a scientist* before one starts making judgements about science itself. So when it comes to matters of social responsibility we feel that our judgements are no better or worse than anyone else's. The question therefore arises: is there anything of use that we can say *as scientists* on this subject?

As it happens there is one area of research which is relevant to these important issues. Professor Stanley Milgram has conducted a number of well-conceived and very revealing experiments on obedience to authority and the behaviour of people working for a cause. These experiments were described by Hilary Roberts, a research psychologist, as follows:

> Milgram's experiments concern the behaviour of people working for a cause. Subjects were told they were taking part in a scientific study to discover whether punishment has a positive effect on the learning process. Their task was to train a fellow volunteer (the learner) to remember pairs of words by giving him an electric shock after every mistake.
>
> The subject was told to increase the shock intensity by 15 volts after each mistake, from 15 to 450 volts. Shocks were delivered whenever the subject pressed a switch on the shock generator. The switches were labelled with their voltages and with inscriptions ranging from Slight Shock to Danger – Severe Shock.
>
> As the shocks became more painful, the learner began to scream and demand to be released. If the subject expressed reluctance to proceed, the experimenter replied:
>
> 'Please go on' or 'The experiment requires that you continue.' At the higher levels of shock the learner refused to answer any more questions, saying he was no longer part of the experiment. The subject was then told to treat no answer as a wrong answer and to continue giving the shocks.
>
> Unknown to the subject, the shock generator was a fake and the learner an actor. The real aim of the experiment was to see if subjects would obey a figure in authority (the experimenter) even when he told them to give painful shocks to an innocent victim.
>
> Out of more than a thousand subjects, over 60 per cent continued to 450 volts. . . . Those who administered the high voltage shocks were not sadists, they were simply ordinary people obeying orders in the cause of science.
>
> Given a suitable cause, the majority of people will carry out orders they would otherwise find unthinkable. They cease to be responsible for their own actions.
>
> *The Guardian*, 1 February 1984.

We feel that these experiments are worth drawing to the attention of scientists and engineers working in areas of research, particularly weapons research, which have evident potential for direct or indirect harm to humanity. They are also relevant to scientists working with animals, for if our ability to inflict pain on other human beings can be so easily manipulated, how much more easily may we become disinterested in the pain we inflict on our fellow mammals?

Any random sample of scientists will contain a broad spectrum of attitudes to these problems of social responsibility. We hope that the majority of scientists will have thought seriously about such issues and made up their own minds as to what is the right course of action for them. To this group we have nothing to say. However, there will always be, in any such sample, a group who is simply not interested in the effects and implications of their work and who wish only to be left alone to get on with their job. To this group we would say: if you have not considered the social implications of your work and decided whether or not your conscience will permit you to do it without reservation; if you have not asked yourself whether your work as a scientist is committing you to activities which you would not engage in as a lay person, then you are not only avoiding your civic responsibilities – you are behaving in a manner which is incompatible with being a good scientist.

Further reading list

Bernal, J.D., *The Extension of Man: A History of Physics before 1900*, Paladin, 1973

Beveridge, W.I.B., *The Art of Scientific Investigation*, W. Heinemann Ltd, 1961

Broad, W. and Wade, N., *Betrayers of the Truth: Fraud and Deceit in the Halls of Science*, Century Publishing Co. Ltd, 1983

Bronowski, J., *The Common Sense of Science*, Heinemann Educational, 1951

Bronowski, J., *The Ascent of Man*, BBC Publications, 1973

Dixon, B., *What is Science For?*, Pelican, 1976

Gardner, M., *Science Good, Bad and Bogus*, Prometheus, 1981

Goodfield, J., *An Imagined World*, Harper and Row, 1981

Harré, R., *Great Scientific Experiments*, Oxford University Press, 1983

Kitcher, P., *Abusing Science: The Case against Creationism*, M.I.T. Press, Cambridge, Massachusetts, 1983

Levinson, H.C., *The Science of Chance*, Faber and Faber, 1967

Meadows, D.H., Meadows, D.L., Randers, J., Behrens, W., *The Limits to Growth*, Universe Books, New York, 1972, 1979 (2nd edition)

Medawar, Sir P.B., *Advice to a Young Scientist*, Harper and Row, 1981

Medawar, Sir P.B., *The Art of the Soluble*, Pelican, 1969

Mesarovic, M., Pestel, E., *Mankind at the Turning Point*, Hutchinson, 1975

Polanyi, M., *Personal Knowledge*, University of Chicago Press, 1958

Sagan, C., *Cosmos*, Macdonald Futura, 1981

Sagan, C., *The Dragons of Eden: Speculations on the Evolution of Human Intelligence*, Coronet Books, Hodder and Stoughton, 1977

Ziman, J., *Public Knowledge*, Cambridge University Press, 1968

Ziman, J., *The Force of Knowledge*, Cambridge University Press, 1976

The Skeptical Inquirer, vol. VI, No. 3, Spring 1982